CW01661942

LIVERPOOL's RAILWAY STATIONS

1830-1985

COMPILED BY ROB GELL

HEYDAY PUBLISHING COMPANY

DEDICATED TO THE MEMORY OF CHARLES E. BOX

Given the boundaries of Seaforth, Aintree, Fazakerley, Broadgreen, Hunts Cross and Speke, how many railway passenger stations has Liverpool hosted? Most people would say about 40 whilst the correct answer is a staggering 85! Liverpudlians nowadays tend only to think of the Southport, Ormskirk and Manchester lines but this is only half the picture. The now gone but legendary Liverpool Overhead Railway contributed 21 alone (but only 17 at any one time). The Cheshire Lines Committee railway from Halewood via West Derby to Aintree finally closed in 1960 but had once been part of that company's main line between Manchester and Southport. Many race-goers from Eastern counties traversed this route on Grand National day. The line between Bootle and Edge Hill which now only handles container traffic once provided commuters with stations such as Spellow (County Road), Tuebrook and Edge Lane. Many other districts boasted a passenger connection including some that the oldest of inhabitants would hardly recall – examples being Gladstone Dock (Lancashire & Yorkshire Railway) which closed in 1924 and Huskisson (Cheshire Lines Committee) whose shutters went up as early as 1885. The following pages table Liverpool's 85 stations since 1830 together with photographs, useful dates and other information. A map showing passenger and goods stations is also included – see centre pages.

Grateful thanks are extended to the following:

P. Andrews	G. Coltas	D. Ibbotson	F. W. Shuttleworth
P. Bolger	Crosby Library	B. Mathews	R. Stacey
A. W. Burges	J. W. Gahan	J. Peden	D. Thompson
H. C. Casserley	P. H. Hanson	Photomatic	M. Turner
R. M. Casserley	B. Hilton	R. C. Riley	P. Woolley
	I. G. Holt	J. Ryan	

All rights reserved. No part of this book may be reproduced, stored in a retrieval system, or transmitted in any form or by any means, electronic, mechanical, photocopying, recording or otherwise without the prior permission of the Publisher in writing.

ISBN 0 947562 03 6

First Published 1985
Copyright Heyday Publishing Company 1985

HEYDAY PUBLISHING COMPANY
KINGSWAY HOUSE, 83 CROSBY ROAD NORTH,
CROSBY, MERSEYSIDE, L22 4QD, ENGLAND.

Aintree Central (Cheshire Lines Committee) *Line opened* 13-7-1880 *Station opened* 13-7-1880 *Station closed* 7-11-1960 but used for race traffic up to March 1963 *Line closed* 7-12-1964 except for private traffic to sidings. Remains are still visible from Park Road (near Parkfield Avenue).

Aintree Central in the 1930's (Photo D. Thompson)

Aintree Sefton Arms (Lancashire & Yorkshire Railway *Line opened* 2-4-1849 *Station opened* 2-4-1849 Still open. Line electrified in 1906. The Westernmost platforms once accommodated an electric service from Gladstone Dock (LYR) via Ford together with through Liverpool Overhead Railway trains on race days.

Aintree Sefton Arms around the turn of the century (Photo B. Mathews Collection)

Aintree Racecourse (Lancashire & Yorkshire Railway) *Line opened* 27-8-1866 *Station opened* circa 1890 *Station closed* 31-3-1962 Line still open but almost nothing remains of the station. It was only used for Grand National excursion traffic and was known as 'Aintree Cinder Lane' up to 18-5-1910. The station was sited on Warbreck Moor near Melling Avenue.

Aintree Racecourse exterior in early days (Photo J. Ryan Collection)

Alexandra Dock (Liverpool Overhead Railway) *Line opened* 4-2-1893 *Station opened* 4-2-1893 *Station closed* 30-12-1956 *Line closed* 30-12-1956 The station was situated on Regent Road between Strand Road and Church Street.

A ground level view of Alexandra Dock (L.O.R.) from within the dock complex after closure. (Photo J. Peden)

Alexandra Dock (London & North Western Railway) *Line opened* 1-1-1880 *Station opened* 1-9-1881 *Station closed* 31-5-1948 Line still open for container traffic but nothing remains of this one platform terminus. It was for situated on Regent Road north of the junction with Nelson Street.

An enthusiasts special at Alexandra Dock (LNWR) in 1959. Photo taken from Derby Road near the old Bootle Hospital (Photo R. M. Casserley)

Allerton (London & North Western Railway) *Line opened* 15-2-1864 *Station opened* 15-2-1864 Still open. Originally double track, the line was quadrupled on 13-7-1891. The station's original name was 'Allerton for Garston'.

Allerton (LNWR) in the 1930's before modernisation. (Photo D. Thompson)

Bank Hall (Lancashire & Yorkshire Railway) *Line opened* 1-10-1850 *Station opened* 1870 Still open. Line electrified in 1904.

Bank Hall Station in 1965. (Photo D. Thompson)

Bootle Balliol Road (London & North Western Railway) *Line opened* 1-1-1880 *Station opened* 1-9-1881 *Station closed* 31-5-1948 Line still open for container traffic. Remains are still visible and the booking hall stands intact facing Bootle Museum entrance in Oriel Road.

Bootle Balliol Road (LNWR) shortly after closure showing the Millers Bridge Roadbridge overhead. (Photo D. Thompson)

Bootle Oriel Road (Lancashire & Yorkshire Railway) *Line opened* 1-10-1850 *Station opened* circa 4-1876 Still open. Line electrified in 1904. This venue replaced 'Bootle Village' and 'Millers Bridge' stations.

Bootle Oriel Road (LYR) circa 1910. (Photo P. Woolley Collection)

Bootle Village (Lancashire & Yorkshire Railway) *Line opened* 1-10-1850 *Station opened* circa 11-1850 *Station closed* circa 4-1876 (replaced by Bootle Oriel Road – see above) Line still open. This early station almost certainly stood on the Northern side of Merton Road.

Breck Road (London & North Western Railway) *Line opened* 15-10-1866 *Station opened* 1-7-1870 *Station closed* 31-5-1948 Line still open for container traffic but all traces of the station (Townsend Lane) have gone.

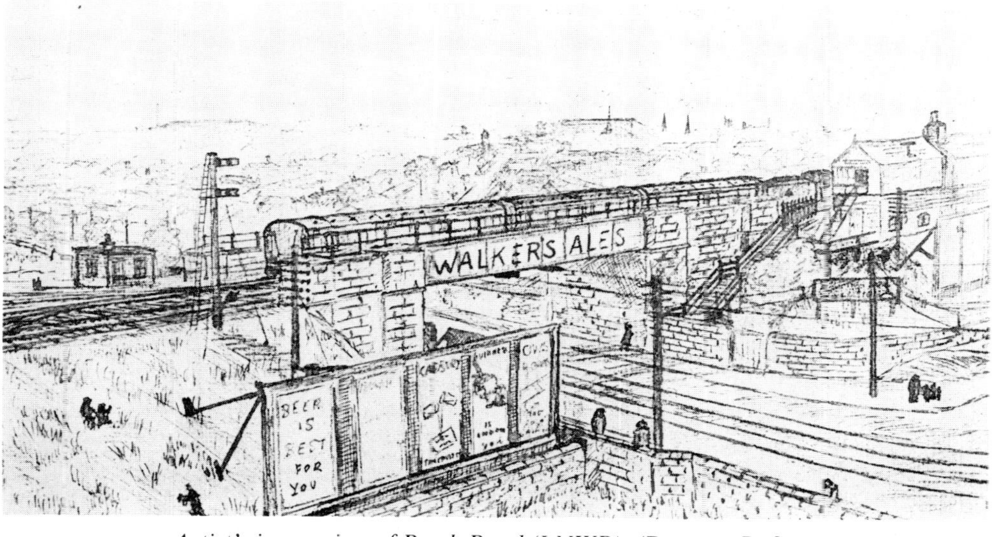

Artist's impression of Breck Road (LNWR). (Drawing R. Stacey)

Broad Green (London & North Western Railway) *Line opened* 15-9-1830 *Station opened* 15-9-1830 Still open. This is the earliest station in Liverpool still operating a passenger service, having once been part of the Liverpool & Manchester Railway's original line (the Edge Hill and Lime Street branch not opening until 1836). At first it was little more than a stopping place being without platforms or proper shelter. The photograph below shows that it blossomed into a grand quadruple line station but today the tracks have been reduced to two once more and the buildings are far more humble.

Broad Green (LNWR) between the Wars. (Photo D. Thompson)

Brocklebank Dock (Liverpool Overhead Railway) *Line opened* 4-2-1893 *Station opened* 4-2-1893 *Station closed* 30-12-1956 *Line closed* 30-12-1956 The station was situated on Regent Road between Seymour Street and Millers Bridge.

Brunswick (Cheshire Lines Committee) *Line opened* 1-6-1864 *Station opened* 1-6-1864 *Station closed* 2-3-1874 *Line closed* 1971 This station was the terminus for trains from South Liverpool and beyond until Liverpool Central (CLC) was opened in 1874, passengers being conveyed by coach from Brunswick to the company's offices in Lower Castle Street. The building was sited at the junction of Sefton and Northumberland Streets. At closure the station became part of the large goods depot which was adjacent.

Brunswick Dock (Liverpool Overhead Railway) *Line opened* 4-2-1893 *Station opened* 4-2-1893 *Station closed* 30-12-1956 *Line closed* 30-12-1956 The station was sited on Sefton Street just south of Hill Street.

Brunswick Dock Station (L.O.R.) just before closing. (Photo J. Peden)

Canada Dock (Liverpool Overhead Railway) *Line opened* 4-2-1893 *Station opened* 4-2-1893 *Station closed* 30-12-1956 *Line closed* 30-12-1956 The station was situated on Regent Road south of Bankfield Street.

Canada Dock (London & North Western Railway) *Line opened* 15-10-1866 *Station opened* 1-7-1870 *Station closed* 5-5-1941 *Line closed* 12-9-1982 This station was named 'Bootle' until 1881. It's closure was a result of enemy action when a nearby canal aqueduct was breached flooding the entire complex. Much of this small terminus remains and the booking hall stands on Derby Road south of the junction with Bankhall Street.

The remains of Canada Dock Station (LNWR) in 1985, virtually intact after 44 years of closure. (Photo author)

Childwall (Cheshire Lines Committee) *Line opened* 1-12-1879 *Station opened* 1-12-1879 *Station closed* 1-1-1931 *Line closed* 8-1975 The station was situated on the North West side of Well Lane but nothing now remains, the trackwork being lifted in 1979.

The remains of Childwall Station circa 1935. (Photo D. Thompson)

Clarence Dock (Liverpool Overhead Railway) *Line opened* 4-2-1893 *Station opened* 4-2-1893 *Station closed* 30-12-1956 *Line closed* 30-12-1956 The station was sited on Regent Road opposite Saltney Street.

A Dingle bound train entering Clarence Dock Station. (Photo Photomatic)

Clubmoor (Cheshire Lines Committee) *Line opened* 1-12-1879 *Station opened* 14-4-1927 *Station closed* 7-11-1960 *Line closed* 8-1975 The station stood on the south side of Broad Lane near New Hall Lane but nothing now remains. The trackwork was lifted in 1979.

Clubmoor Station (CLC) circa 1929. (Photo D. Thompson)

Cressington & Grassendale (Cheshire Lines Committee) *Line opened* 1-6-1864 *Station opened* 1-3-1873 *Station closed* 17-4-1972 It re-opened as part of the 'Merseyrail' network on 3-1-1978 and is still open. Since then the name has been 'Cressington' as it was prior to 1877.

Cressington & Grassendale Station (CLC) in 1971. (Photo D. Thompson)

Crown Street (London & North Western Railway) *Line opened* 15-9-1830 *Station opened* 15-9-1830 *Station closed* 15-8-1836 *Line closed* 1-5-1972 This was the much celebrated Western terminus of the Liverpool & Manchester Railway prior to the opening of Lime Street. From Edge Hill the line was on a rising gradient to Crown Street through a narrow tunnel. Coaches were detached from the locos at Edge Hill Moorish Arch (near Chatsworth Street) and hauled up the incline by means of a stationary engine within the Arch and ropes and pulleys. After passenger closure it became devoted to goods and locos were not allowed to work through the tunnel until much later. The station was situated on Crown Street near to the junction with Myrtle Street.

Custom House (Liverpool Overhead Railway) *Line opened* 4-2-1893 *Station opened* 4-2-1893 *Station closed* 30-12-1956 *Line closed* 30-12-1956 The station was renamed 'Canning Dock' in 1947 and was situated at the junction of Strand Street and Canning Place.

Custom House after renaming to 'Canning Dock'. (Photo J. Peden)

Dingle (Liverpool Overhead Railway) *Line opened* 21-12-1896 *Station opened* 21-12-1896 *Station closed* 30-12-1956 *Line closed* 30-12-1956 In 1896 the Overhead Railway extended the South end of the line inland to Dingle. The higher land resulted in the terminus being underground thus belying the railway's name! It was sited on Park Road north of Kedleston Street.

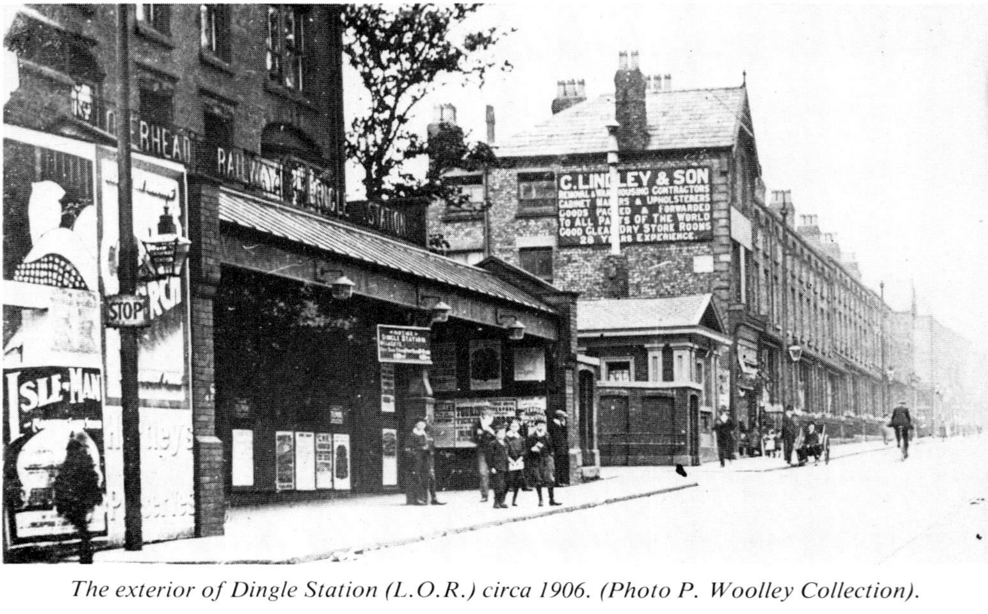

The exterior of Dingle Station (L.O.R.) circa 1906. (Photo P. Woolley Collection).

Edge Hill (London & North Western Railway) *Line opened* 15-8-1836 *Station opened* 15-8-1836 Still open. This is Liverpool's oldest passenger station with surviving original buildings, ex Liverpool & Manchester Railway in part. When passenger traffic became too much for Crown Street, a branch to Lime Street was opened through a twin-track tunnel and Edge Hill was sited between its Eastern portal and the junction with the original line. It was known as 'New Tunnel' station in early days and, as with the Crown Street incline, stationary engines were employed. Coaches to Lime Street were allowed down the gradient by means of gravity, being hauled in the reverse direction by Edge Hill's stationary power. Rope working lasted until 1870.

A 1959 express entering Edge Hill ex Lime Street. (Photo P. H. Hanson)

Edge Lane (London & North Western Railway) *Line opened* 1-6-1866 *Station opened* 11-1870 *Station closed* 31-5-1948 Line still open for container traffic. The station was sited on Edge Lane near Binns Road.

A southbound goods passes Edge Lane (LNWR) in 1948. (Photo D. Thompson)

Fazakerley (Lancashire & Yorkshire Railway) *Line opened* 20-11-1848 *Station opened* 20-11-1848 Still open and now forms part of the 'Merseyrail' network. The station was named 'Simonswood' originally, 'Aintree' by 1850 becoming 'Fazakerley' in 1860.

Fazakerley circa 1910. (Photo P. Bolger Collection)

Ford (Lancashire & Yorkshire Railway) *Line opened* 27-8-1866 *Station opened* 1-6-1906 *Station closed* 2-4-1951 Line still open for goods. See 'Linacre Road' for an outline of the line's electrified services. The station stood on Netherton Way near Bridle Road but no traces remain.

Garston (Cheshire Lines Committee) *Line opened* 1-3-1873 *Station opened* 1-4-1874 *Station closed* 17-4-1972 Re-opened as part of the 'Merseyrail' network on 3-1-1978 Still open.

Garston (CLC) interior in 1954. (Photo H. C. Casserley)

Garston Church Road (London & North Western Railway) *Line opened* 1-7-1852 *Station opened* 1-3-1881 *Station closed* 15-4-1917 *Station re-opened* 5-5-1919 *Station closed* 3-7-1939 Line still open for goods. As the name implies this station was at Church Road on the East side.

Garston Church Road in the 1930's. (Photo D. Thompson)

Garston Dock (London & North Western Railway) *Line opened* 1-7-1852 *Station opened* 1-7-1852 *Station closed* 15-4-1917 *Station re-opened* 5-5-1919 *Station closed* 16-6-1947 Line still open for goods but the station area has been reclaimed. It stood at the North end of Dock Road, east side.

Garston Dock (LNWR) in the 1930's. (Photo D. Thompson)

Gateacre (Cheshire Lines Committee) *Line opened* 1-12-1879 *Station opened* 1-12-1879 *Station closed* 17-4-1972 *Line closed* 8-1975 The station was sited on the North side of Belle Vale Road, the tracks being lifted in 1979.

Gateacre Station (CLC) in Edwardian Days. (Photo P. Bolger Collection)

Gladstone Dock (Lancashire & Yorkshire Railway) *Line opened* 27-8-1866 *Station opened* 7-9-1914 *Station closed* 7-7-1924 *Line closed* 10-6-1968 This elevated single platform station was on the West side of Regent Road about 250 yards from the junction with Rimrose Road and should not be confused with the Liverpool Overhead station of the same name (opened 1930) – see next page.

Gladstone Dock (LYR) in 1914. Note the trams on the right – today, buses still terminate at this point (Regent Road/Rimrose Road). Note also the Overhead Railway's Seaforth Sands Station in the distance. (Photo Authors Collection)

Gladstone Dock (Liverpool Overhead Railway) *Line opened* 4-2-1893 *Station opened* 16-6-1930 *Station closed* 30-12-1956 *Line closed* 30-12-1956 This station overlooked North Mersey goods yard (ex LYR) and the nearest road junction was Regent Road and Grove Street.

Gladstone Dock (L.O.R.). (Photographer Not Known)

Great Howard Street (Lancashire & Yorkshire Railway) *Line opened* 20-11-1848 *Station opened* 20-11-1848 *Station closed* 13-5-1850 *Line closed* 30-9-1963 This was the original Liverpool terminus of the Lancashire & Yorkshire and East Lancashire Railways prior to the opening of Tithebarn Street station further South on 13-5-1850 (see Liverpool Exchange). Upon closure the site was entirely given up for goods use. As the name implies, it was situated on Great Howard Street and was close to Formby Street.

Herculaneum Dock (1st. Station) (Liverpool Overhead Railway) *Line opened* 4-2-1893 *Station opened* 4-2-1893 *Station closed* 20-12-1896 *Line closed* 30-12-1956 This was the original Southern terminus of the Overhead Railway being replaced by a second station of the same name in 1895 when the Southern extension to Dingle was brought into use. The first station became used as a carriage shed from this date. It stood in dockland where projections of Sefton Street and Beresford Road would have met (see overleaf for 2nd Station).

The 1st Station at Herculaneum Dock in use as a Carriage Shed. (Photo Photomatic)

Herculaneum Dock (2nd Station) (Liverpool Overhead Railway) *Line opened* 4-2-1893 *Station opened* 21-12-1896 *Station closed* 30-12-1956 *Line closed* 30-12-1956 This station replaced the former Herculaneum Dock station (see previous page) and stood 165 yards further north.

Hunts Cross (Cheshire Lines Committee) *Line opened* 1-3-1873 *Station opened* 5-1874 Still open. Originally double track, the station was rebuilt in 1883 to accommodate the quadrupling of lines only to revert back to main-line double track in recent years.

Hunts Cross (CLC) between the Wars with four platforms evident. (Photo D. Thompson)

Huskisson (Cheshire Lines Committee) *Line opened* 1-7-1880 *Station opened* 13-7-1880 *Station closed* 1-5-1885 *Line closed* 8-1975 This long forgotten terminus was situated in the expansive goods yard of the same name. The station building adjoined Boundary Street just east of its junction with Steel Street but nothing now remains, the tracks having been lifted from the whole area.

The remains of Huskisson (CLC) in 1954. (Photo D. Thompson)

Huskisson Dock (Liverpool Overhead Railway) *Line opened* 4-2-1893 *Station opened* 1896 *Station closed* 30-12-1956 *Line closed* 30-12-1956 This station was situated on Regent Road just south of Sandhills Lane.

A train entering Huskisson Dock (L.O.R.) in the 1950's. (Photo Photomatic)

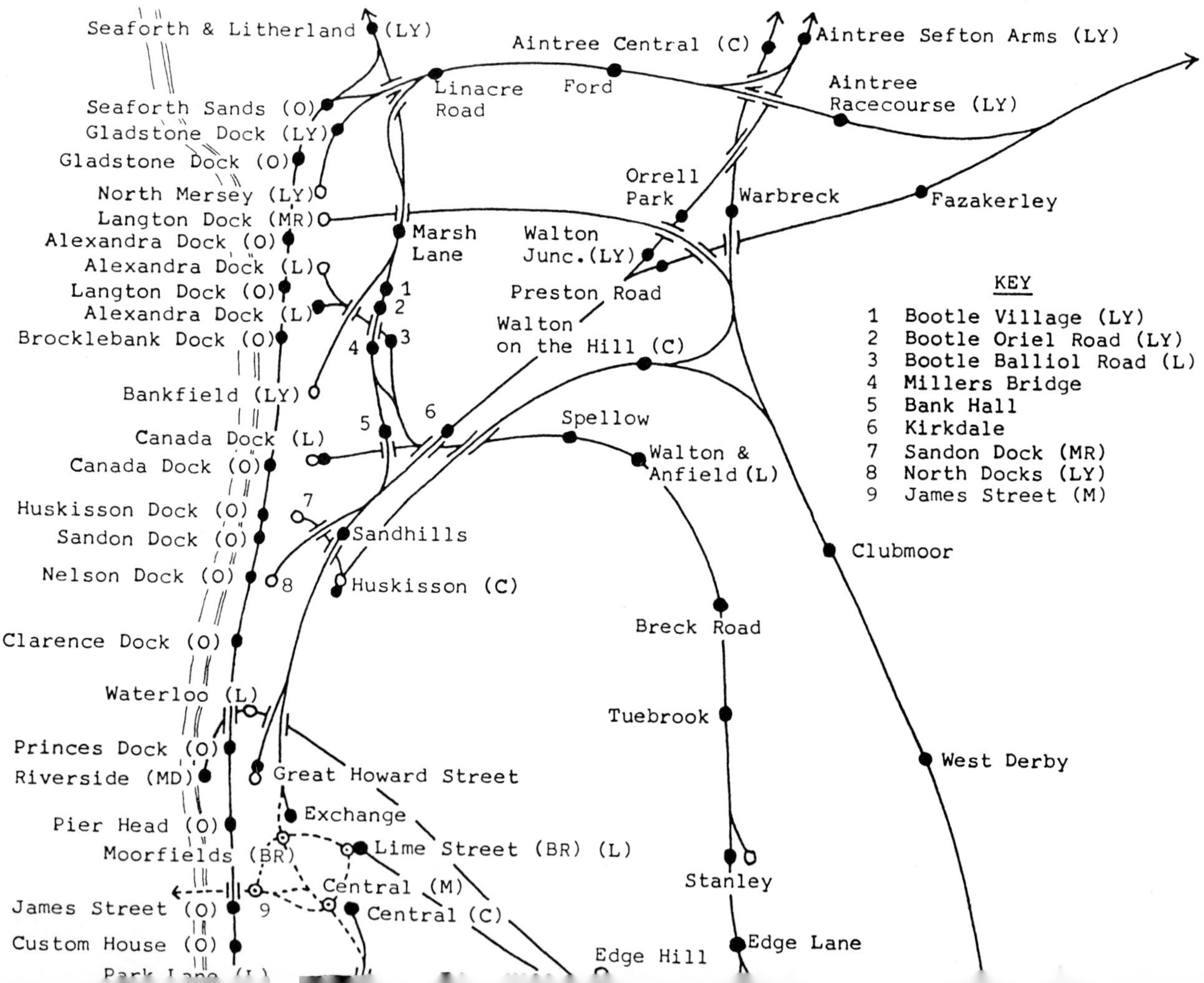

Seaforth & Litherland (LY)
Aintree Central (C)
Aintree Sefton Arms (LY)
Aintree Racecourse (LY)
Linacre Road
Ford
Seaforth Sands (O)
Gladstone Dock (LY)
Gladstone Dock (O)
North Mersey (LY)
Langton Dock (MR)
Alexandra Dock (O)
Alexandra Dock (L)
Langton Dock (O)
Alexandra Dock (L)
Brocklebank Dock (O)
Bankfield (LY)
Canada Dock (L)
Canada Dock (O)
Huskisson Dock (O)
Sandon Dock (O)
Nelson Dock (O)
Clarence Dock (O)
Waterloo (L)
Princes Dock (O)
Riverside (MD)
Pier Head (O)
Moorfields (BR)
James Street (O)
Custom House (O)
Orrell Park
Warbreck
Fazakerley
Marsh Lane
Walton Junc.(LY)
Preston Road
Walton on the Hill (C)
Spellow
Walton & Anfield (L)
Sandhills
Huskisson (C)
Great Howard Street
Exchange
Lime Street (BR) (L)
Central (M)
Central (C)
Edge Hill
Breck Road
Tuebrook
Stanley
Edge Lane
Clubmoor
West Derby
KEY
1 Bootle Village (LY)
2 Bootle Oriel Road (LY)
3 Bootle Balliol Road (L)
4 Millers Bridge
5 Bank Hall
6 Kirkdale
7 Sandon Dock (MR)
8 North Docks (LY)
9 James Street (M)

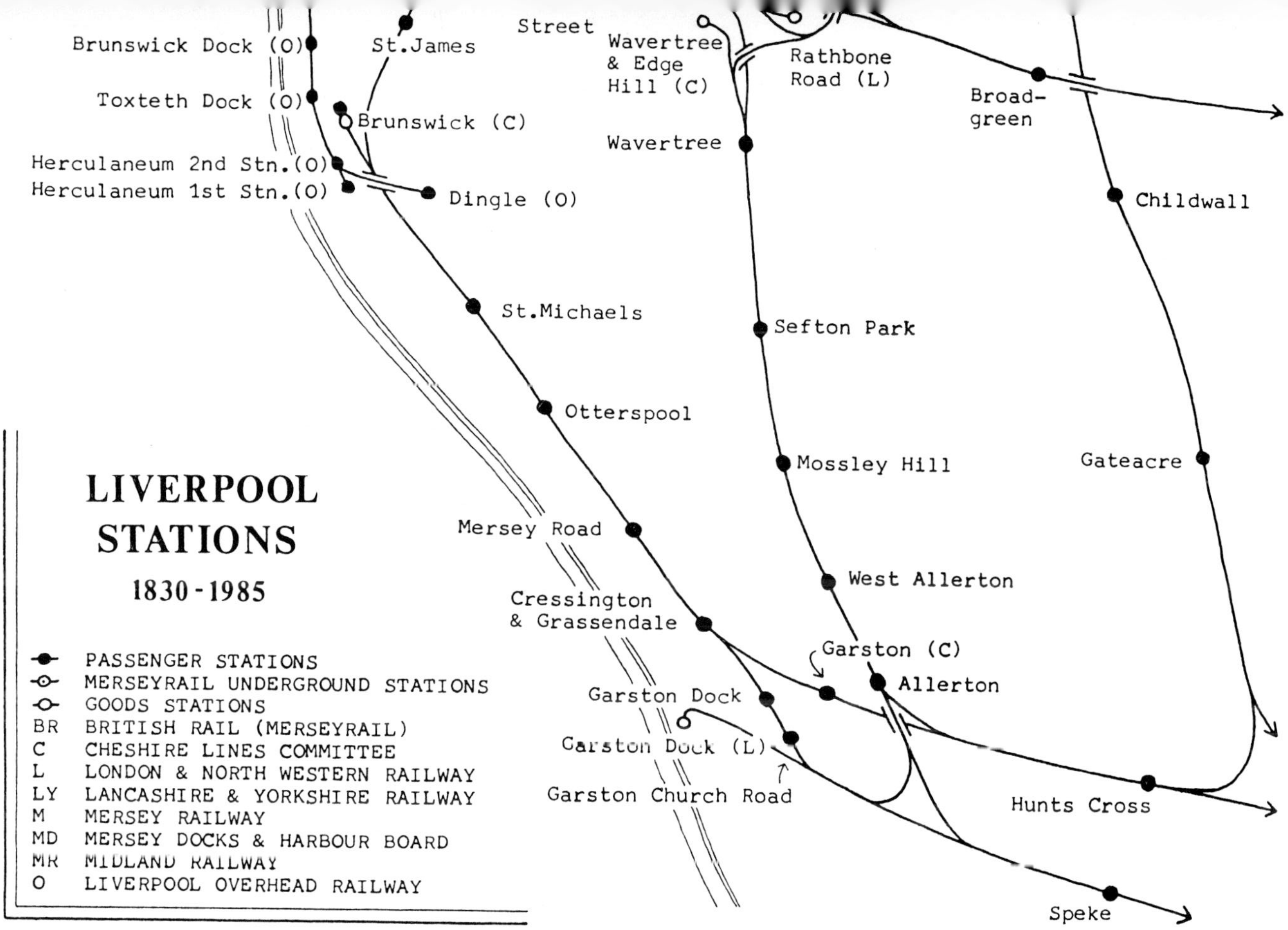
LIVERPOOL STATIONS
1830-1985
PASSENGER STATIONS
MERSEYRAIL UNDERGROUND STATIONS
GOODS STATIONS
BR BRITISH RAIL (MERSEYRAIL)
C CHESHIRE LINES COMMITTEE
L LONDON & NORTH WESTERN RAILWAY
LY LANCASHIRE & YORKSHIRE RAILWAY
M MERSEY RAILWAY
MD MERSEY DOCKS & HARBOUR BOARD
MR MIDLAND RAILWAY
O LIVERPOOL OVERHEAD RAILWAY
Brunswick Dock (O)
Toxteth Dock (O)
Herculaneum 2nd Stn.(O)
Herculaneum 1st Stn.(O)
St.James
Brunswick (C)
Dingle (O)
Street
Wavertree & Edge Hill (C)
Rathbone Road (L)
Broad-green
Wavertree
Childwall
St.Michaels
Sefton Park
Otterspool
Mossley Hill
Gateacre
Mersey Road
West Allerton
Cressington & Grassendale
Garston (C)
Allerton
Garston Dock
Garston Dock (L)
Garston Church Road
Hunts Cross
Speke

James Street (Liverpool Overhead Railway) *Line opened* 4-2-1893 *Station opened* 4-2-1893 *Station closed* 30-12-1956 *Line closed* 30-12-1956 The station was situated on Strand Street between James and Redcross Streets.

James Street (L.O.R.). (Photographer Not Known)

James Street (Mersey Railway) *Line opened* 20-1-1886 *Station opened* 20-1-1886 Still open. The line was electrified in 1903 largely to combat the foul smell resulting from the effects of smoke upon the sulphurous rock within the tunnels. This station was a terminus until 1892 when the line was extended to Liverpool Central (Mersey Railway).

James Street (Mersey Railway) in the 1960's. (Photo Authors Collection)

Kirkdale (Lancashire & Yorkshire Railway) *Line opened* 20-11-1848 *Station opened* 20-11-1848 Still open. The station was known as 'Bootle Lane' until 1876 and the line was electrified in 1906.

Kirkdale Station (LYR) between the Wars. (Photo D. Thompson)

Knotty Ash (Cheshire Lines Committee) *Line opened* 1-12-1879 *Station opened* 1-12-1879 *Station closed* 7-11-1960 *Line closed* 8-1975 The proper title of the station was 'Knotty Ash & Stanley' but prior to 1888 it was named 'Old Swan & Knotty Ash'. Although the tracks were lifted in 1979 traces of the station can still be seen near East Prescot & Chatterton Roads.

Knotty Ash Station in 1954 showing the East Prescot Road Bridge in the distance. (Photo H. C. Casserley)

Langton Dock (Liverpool Overhead Railway) *Line opened* 4-2-1893 *Station opened* circa 1896 *Station closed* 1906 *Line closed* 30-12-1956 The station was situated on Regent Road 185 yards north of Nelson Street.

Linacre Road (Lancashire & Yorkshire Railway) *Line opened* 27-8-1866 *Station opened* 1-6-1906 *Station closed* 2-4-1951 Line still open for goods. As with Ford station, this venue saw electrified traffic between Marsh Lane and Aintree; Gladstone Dock (LYR) and Aintree together with through Liverpool Overhead Railway services (via Seaforth Sands), the latter operating on Grand National days. As the name implies, the station stood on Linacre Road alongside the bus garage. It has long been dismantled but the stumps of the platform supports can still be detected when viewed from Hawthorne Road bridge (near the Tar Distillery).

Liverpool Central (Cheshire Lines Committee) *Line opened* 2-3-1874 *Station opened* 2-3-1874 *Station closed* 17-4-1972 *Line closed* 17-4-1972 The site of this important terminus is now occupied by the new Ranelagh Street shopping complex.

Liverpool Central (CLC) interior in 1953. (Photo Authors Collection)

Liverpool Central (Mersey Railway) *Line opened* 11-1-1892 *Station opened* 11-1-1892 *Station closed* 28-7-1975 Line electrified in 1903 – see James Street (Mersey Railway) also. *Station re-opened* in 1977 with two levels as part of the 'Merseyrail' network. Still open. Deep Level opened 9-5-1977 for James Street loop services whilst Low Level opened 2-5-1977 hosting trains on the Southport, Ormskirk and Kirkby lines – Garston trains began running on 3-1-1978. The original station was a terminus only.

The exterior of both Central Stations circa 1910. The Mersey Railway entrance can be seen right of centre. (Photo Heyday Publishing Company)

Liverpool Exchange (Lancashire & Yorkshire Railway) *Line opened* 13-5-1850 *Station opened* 13-5-1850 *Station closed* 30-4-1977 Line electrified in 1904. Line still open but bypasses station site and descends underground to Moorfields Low Level – see centre pages. Much of the station has been redeveloped as a car park. Originally the station on this site was called 'Tithebarn Street', the track level being 19 feet higher. Extensive rebuilding in 1886 resulted in the lowering of the terminus and it re-opened as Liverpool Exchange on the 12th December of that year.

Liverpool Exchange exterior between the Wars. (Photo Heyday Publishing Company)

Liverpool Lime Street (British Rail) *Line opened* 9-5-1977 *Station opened* 30-10-1977 Still open. This station is part of the deep level Merseyrail loop service between Moorfields and Liverpool Central.

Liverpool Lime Street (Deep Level) in 1985. (Photo Author)

Liverpool Lime Street (London & North Western Railway) *Line opened* 15-8-1836 *Station opened* 15-8-1836 Still open. This station replaced Crown Street as the Liverpool & Manchester Railway's passenger terminus. At first there were only two platforms and trains were lowered from Edge Hill by rope until March 1870 when engines were allowed to work through the tunnel (see Edge Hill also). Most of the tunnel was opened out into cuttings by 1881 and the line quadrupled in 1885.

Liverpool Lime Street (LNWR) interior in 1974. (Photo G. Coltas)

Liverpool Riverside (Mersey Docks & Harbour Board) *Line opened* 12-6-1895 *Station opened* 12-6-1895 *Station closed* 25-2-1971 *Line closed* 25-2-1971 This station catered for the passenger traffic to and from the ocean going liners that once graced the Mersey. It was sited in dockland and was connected with the main-line network via dock lines to Waterloo Goods station (LNWR) where Edge Hill could then be reached. The station was situated between Princes Dock and the landing stage.

Riverside Station in early days. (Photo J. Ryan Collection)

Marsh Lane & Strand Road (Lancashire & Yorkshire Railway) *Line opened* 1-10-1850 *Station opened* circa 11-1850 Still open. The station was known simply as 'Marsh Lane' prior to 1886. In 1966 its name was changed to 'Bootle New Strand' in recognition of the large shopping complex built nearby. The line was electrified in 1904.

Marsh Lane & Strand Road in 1965. (Photo D. Thompson)

Mersey Road & Aigburth (Cheshire Lines Committee) *Line opened* 1-6-1864 *Station opened* 1-6-1864 *Station closed* 17-4-1972 *Station re-opened* 3-1-1978 as part of the 'Merseyrail' network. Still open. It was known simply as 'Mersey Road' prior to 1880 and was renamed 'Aigburth' upon re-opening.

Mersey Road & Aigburth Station in early days. (Photo Heyday Publishing Company)

Millers Bridge (Lancashire & Yorkshire Railway) *Line opened* 1-10-1850 *Station opened* 12-1850 *Station closed* circa 4-1876 (replaced by Bootle Oriel Road – see entry). Line still open. As the name implies the station stood by Millers Bridge but no trace remains today.

Moorfields (British Rail) Deep Level *Line opened* 9-5-1977 *Station opened* 6-5-1978 Low Level *Line opened* 2-5-1977 *Station opened* 2-5-1977 Both still open and form part of the 'Merseyrail' network.

Moorfields Station exterior in 1985. (Photo Author)

Mossley Hill (London & North Western Railway) *Line opened* 15-2-1864 *Station opened* 15-2-1864 Still open. In 1891 the line was quadrupled and the station site was moved further north. The station was more formally named 'Mossley Hill for Aigburth'.

Mossley Hill Station between the Wars. (Photo D. Thompson)

Nelson Dock (Liverpool Overhead Railway) *Line opened* 4-2-1893 *Station opened* 1896 *Station closed* 30-12-1956 *Line closed* 30-12-1956 The station stood on Regent Road near the junction with Fulton Street.

Orrell Park (Lancashire & Yorkshire Railway) *Line opened* 2-4-1849 *Station opened* circa 1906 Still open. The line was electrified in 1906.

Orrell Park Station when new. (Photo P. Bolger Collection)

Otterspool (Cheshire Lines Committee) *Line opened* 1-6-1864 *Station opened* 1-6-1864 *Station closed* 5-3-1951 Line still open as part of the 'Merseyrail' network. The station was known as 'Otter's Pool' prior to 1866.

Otterspool Station (CLC) circa 1910. (Photo P. Bolger Collection)

Pier Head (Liverpool Overhead Railway) *Line opened* 4-2-1893 *Station opened* 4-2-1893 *Station closed* 30-12-1956 *Line closed* 30-12-1956 This, the busiest station on the Overhead, was situated alongside the Royal Liver building.

The Overhead Railway & Pier Head Station in 1908. The Old Georges Dock can be seen making way for the coming of the Liver Building (opened in 1911). (Photo Authors Collection)

Preston Road (Lancashire & Yorkshire Railway) *Line opened* 20-11-1848 *Station opened* 20-11-1848 Still open as part of the 'Merseyrail' network and was renamed 'Rice Lane' on 14-5-1984.

An Edwardian view of Preston Road (LYR) (Photo Authors Collection)

Princes Dock (Liverpool Overhead Railway) *Line opened* 4-2-1893 *Station opened* 4-2-1893 *Station closed* 30-12-1956 *Line closed* 30-12-1956 This station was situated on Waterloo Road at the junction with Roberts Street.

St. James (Cheshire Lines Committee) *Line opened* 2-3-1874 *Station opened* 2-3-1874 *Station closed* 1-12-1916 Line still open as part of the 'Merseyrail' network. The station was in a deep cutting under Ashwell Street.

St. James platforms circa 1935. (Photo D. Ibbotson)

St. Michaels (Cheshire Lines Committee) *Line opened* 1-6-1864 *Station opened* 1-6-1864 *Station closed* 17-4-1972 *Station re-opened* 3-1-1978 as part of the 'Merseyrail' network. Still open.

A local train at St. Michaels Station in 1963. (Photo I: G. Holt)

Sandhills (Lancashire & Yorkshire Railway) *Line opened* 20-11-1848 *Station opened* 6-1854 Still open. Line electrified in 1904. This station was named 'North Docks' prior to 1857.

Sandhills Station in 1965. (Photo D. Thompson)

Sandon Dock (Liverpool Overhead Railway) *Line opened* 4-2-1893 *Station opened* 4-2-1893 *Station closed* circa 1896 *Line closed* 30-12-1956 This station was replaced by Huskisson Dock and Nelson Dock stations which opened North and South of it respectively. Sandon Dock station was situated on Regent Road equidistant between Boundary Street and Sandhills Lane.

Seaforth & Litherland (Lancashire & Yorkshire Railway) *Line opened* 1-10-1850 *Station opened* 1-10-1850 Still open. Line electrified in 1904. The station was known simply as 'Seaforth' prior to 1905.

Seaforth & Litherland station in 1954. A Liverpool Overhead Railway electric train can just be seen to the left of the nameboard. (Photo D. Thompson)

Seaforth Sands (1st Station) (Liverpool Overhead Railway) *Line opened* 30-4-1894 *Station opened* 30-4-1894 *Station closed* 1925 *Line closed* 30-12-1956 After closure this station became a carriage shed and, as such, served its successor, the 2nd station (see overleaf) until complete closure of the railway. It should be noted that both stations hosted passengers between 1905 & 1925. The 1st station was situated directly opposite the 'Caradoc' public house on Crosby Road South.

The 1st Station at Seaforth Sands, circa 1900. (Photo Crosby Library)

Seaforth Sands (2nd Station) (Liverpool Overhead Railway) *Line opened* 2-7-1905 *Station opened* 2-7-1905 *Station closed* 30-12-1956 *Line closed* 30-12-1956 This station, built alongside the 1st (see preceeding page), did not take over all the traffic right away as both operated services between 1905 and 1925. The 2nd station stood west of the junction between Regent Road and Crosby Road South.

Seaforth Sands (2nd station) (Photographer Not Known)

Sefton Park (London & North Western Railway) *Line opened* 15-2-1864 *Station opened* 6-1892 *Station closed* 2-5-1960 Line still open. The station stood on Smithdown Road near to the junction with Garmoyle Road.

Sefton Park Station in 1958. (Photo D. Thompson)

Speke (London & North Western Railway) *Line opened* 1-7-1852 *Station opened* 1-7-1852 *Station closed* 22-9-1930 Line still open. The station stood on Woodend Lane (now called Woodend Avenue).

Speke Station in early days. (Photo Authors Collection)

Spellow (London & North Western Railway) *Line opened* 15-10-1866 *Station opened* 9-1882 *Station closed* 31-5-1948 Line still open for container traffic. The station lay beneath County Road and the building can still be seen between Romley and Arnot Streets although the lower frontage has been adapted for business use.

A goods loco passing the remains of Spellow Station in 1958 (Photo J. Peden)

Stanley (London & North Western Railway) *Line opened* 1-6-1866 *Station opened* 1-6-1866 *Station closed* 31-5-1948 Line still open for container traffic. The station was on Prescot Road just East of its junction with Prescot Drive and the buildings were demolished only recently.

Stanley station between the Wars. (Photo D. Thompson)

Toxteth Dock (Liverpool Overhead Railway) *Line opened* 4-2-1893 *Station opened* 4-2-1893 *Station closed* 30-12-1956 *Line closed* 30-12-1956 The station was situated on Sefton Street south of Park Street.

Tuebrook (London & North Western Railway) *Line opened* 1-6-1866 *Station opened* 1-6-1866 *Station closed* 31-5-1948 Line still open for container traffic. The station was sited on the north side of West Derby Road between Dorset and Windsor Roads.

Tuebrook Station in the 1930's, looking south. (Photo D. Thompson)

Walton & Anfield (London & North Western Railway) *Line opened* 15-10-1866 *Station opened* 1-7-1870 *Station closed* 31-5-1948 Line still open for container traffic. The station was situated on the East Side of Walton Lane at the junction with Church Road. Numerous Goodison Park football specials were accommodated here and prior to 1910 it was named 'Walton for Aintree'. Platform remains are still visible at this site.

Walton (LNWR) with original name. (Photo Heyday Publishing Company)

Walton Junction (Lancashire & Yorkshire Railway) *Line opened* 2-4-1849 *Station opened* 2-4-1849 Still open. Line electrified in 1906.

Walton Junction (LYR) in the 1930's. (Photo D. Thompson)

Walton-on-the-Hill (Cheshire Lines Committee) *Line opened* 1-12-1879 *Station opened* 1-12-1879 (as a terminus – the line to Huskisson was not opened to passengers until 13-7-1880) *Station closed* 1-1-1918 *Line closed* 8-1975 The station stood north east of the Rice Lane and Queens Drive junction and remains could be seen up till a few years ago. Today, a housing development occupies the land.

Walton-on-the-Hill's remains in the 1960's. (Photo D. Thompson)

Wapping Dock (Liverpool Overhead Railway) *Line opened* 4-2-1893 *Station opened* 4-2-1893 *Station closed* 30-12-1956 *Line closed* 30-12-1956 The station was situated on Chaloner Street at the junction with Blundell Street.

Wapping Dock Station in the 1950's. (Photo A. W. Burges)

Warbreck (Cheshire Lines Committee) *Line opened* 13-7-1880 *Station opened* 1-8-1929 *Station closed* 7-11-1960 *Line closed* 7-12-1964 (except for private traffic to sidings at Aintree). The line was lifted in 1979. The station stood south of the junction at Walton Vale and Cedar Road.

Warbreck Station in the 1930's. (Photo D. Thompson)

Wavertree (London & North Western Railway) *Line opened* 15-2-1864 *Station opened* 9-1870 *Station closed* 5-8-1958 Line still open. The station stood on the south side of Wellington Road opposite the junction with Nebo Street.

Wavertree station in the year of closure, 1958. (Photo D. Thompson)

Wavertree Lane (London & North Western Railway) *Line opened* 15-9-1830 *Station opened* 15-9-1830 *Station closed* circa 1836 Line still open. Along with Crown Street and Broad Green this was part of the Liverpool & Manchester Railway's original route. It was little more than a stopping place without any platforms or proper shelter but it did have a level crossing for the Lane in question. It most probably closed with the opening of Edge Hill on 15-8-1836 and the site (along with Wavertree Lane itself) has long been lost to the jungle of surrounding trackwork. It was situated near to the junction of Picton Road and Spofforth Road.

West Allerton (London & North Western Railway) *Line opened* 15-2-1864 *Station opened* 2-1-1939 Still open.

A passenger train passing through West Allerton in the early 1960's. (Photo I. G. Holt)

West Derby (Cheshire Lines Committee) *Line opened* 1-12-1879 *Station opened* 1-12-1879 *Station closed* 7-11-1960 *Line closed* 8-1975 The tracks were lifted in 1979 but the station building still stands on Mill Lane, Liverpool 12, South West of the junction with Barnfield Drive.

West Derby Station in the 1930's. (Photo D. Thompson)